The Story of Science

The Life Stories of Stars

by Roy A. Gallant

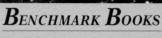

BENCHMARK BOOKS

MARSHALL CAVENDISH
NEW YORK

Series Editor: Roy A. Gallant

Series Consultants:

LIFE SCIENCES
Dr. Edward J. Kormondy
Chancellor and Professor of Biology (retired)
University of Hawaii—Hilo/West Oahu

PHYSICAL SCIENCES
Dr. Jerry LaSala
Department of Physics
University of Southern Maine

Benchmark Books
Marshall Cavendish Corporation
99 White Plains Road
Tarrytown, NY 10591-9001

Library of Congress Cataloging-in-Publication Data
Gallant, Roy A.
        The life stories of stars / by Roy A. Gallant.
        p. cm. — (The story of science)
Includes bibliographical references and index.
Summary: Describes what was believed in the past about stars, including the sun, and what we know today.
ISBN 0-7614-1152-6
        1. Stars—juvenile literature. 2. Astronomy—History—Juvenile literature. [1. Stars. 2. Astronomy.] I. Title. II. Series.
QB801.7.G34   2000   99-086675   CIP   AC

Photo research by Linda Sykes
Diagrams on pp. 14, 30, 34, 38, 40, 45, 48, 52, 54, 70, by Jeannine L. Dickey

Cover photo: *Hubble Space Telescope image of "giant twisters" in a great cloud of space gas and dust called the Lagoon Nebula, some 5,000 light-yars away.* NASA

Photo credits: Cover Page- (Lagoon Nebula) Material created with support to AURA/STScI from NASA contract NAS5-26555
Title Page, 6 (star field), 49, 63, 64, 65 (cygnus loop), 68, 69- Material created with support to AURA/STScI from NASA contract NAS5-26555; 8- Courtesy of SOHO consortium. SOHO is a project of international cooperation between ESA and NASA; 11, 20- www.arttoday.com; 14, 30, 34, 38, 40, 45, 48, 52, 54, 70- Copyright © 2000 Jeannine Dickey and its licensors. All rights reserved; 16,17- Digital Imagery® copyright 2000 PhotoDisc, Inc.; 25, 26, 28, 29, 41, 44, 59- The Granger Collection, New York ; 33- Courtesy Ronan Picture Library; 37- www.arttoday.com & J. Dickey; 42- UPI/Corbis-Bettmann; 56- Jason Ware; 61, 66- Photo Researchers/Celestial Image Co./Science Photo Library; 61- (Crab Nebula) Image courtesy Paul Scowen, Jeff Hester, and the Mr. Palomar Obeservatories; 71- Material created with support to AURA/ST ScI from NASA contract NAS5-26555 & J. Dickey.

Printed in Hong Kong
6 5 4 3 2 1

For Alex Rossovsky

Why did not somebody teach me the constellations,
and make me at home in the starry heavens,
which are always overhead,
and which I don't half know to this day?

—Thomas Carlyle

Contents

One

Stars: Abodes of the Gods

Several thousand years ago the night sky looked just about the same as we see it tonight. The stars appeared only as pinpoints of light, just as they do now. Even through a powerful telescope a star still is only a pinpoint. Only one star appears as a large disk, and that is our local star, the Sun.

As you gaze up at the night sky, once your eyes get used to the dark, you begin to see more and more stars. In all, you can see about 9,000 without using a telescope or binoculars. When your eyes become *dark-adapted*, you begin to see that some stars shine with a reddish light, such as Betelgeuse and Bellatrix in the constellation Orion the Hunter. Others appear yellow, such as the Sun, and Capella in the constellation the Charioteer. Still others

A truly dark sky reveals about 9,000 stars that we can see without a telescope, and those stars have changed hardly at all since the first humans gazed at them thousands of years ago. This view shows what the Hubble Space Telescope sees when pointed toward the Large Magellanic Cloud, one of the Milky Way's satellite galaxies more than 160,000 light-years away.

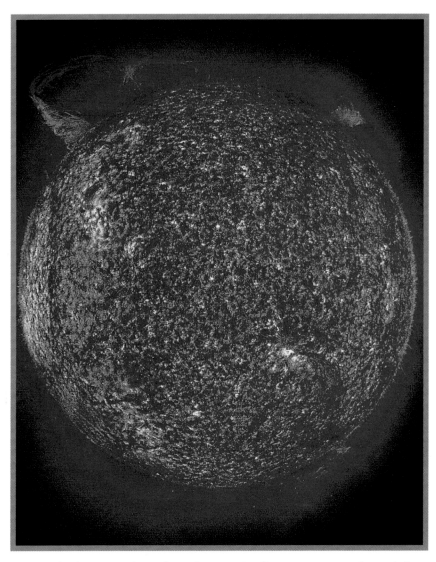

Unlike all other stars, through a telescope the Sun appears as a large disk, and many of its surface details can be seen. At the top are two huge gas eruptions called prominences leaping hundreds of thousands of miles skyward. The hair-like spines seen around the rim are surges of hot gas called spicules. The mottled effect is caused by cells of hot gases welling up from beneath the surface, cooling, and so appearing darker than the surrounding and hotter gases. Today the Sun is some thirty percent hotter than it was when Earth was a young planet.

are blue, such as Rigel in Orion and Sirius in the constellation Canis Major.

One other thing you notice is that the stars parade across the sky from east to west as a group, like soldiers on parade. Because they are so very far away, in a lifetime we cannot notice the stars' real motion as they move this way and that in relation to each other. But if you could watch a nearby star, such as Arcturus in the constellation Bootes, for a thousand years, you would see that it actually does move. Over that period of time, it would move across the sky a distance of half a degree, which is greater than the width of the full moon.

Over the past hundred or more years we have learned much about the stars. We have learned to measure their distance and estimate their size. We have also found out what they are made of, how much matter they contain, how hot they are, how long a star can shine before it dies, and what it's like deep inside. But that knowledge has come slowly, and only with the invention of telescopes and other special instruments that let us examine the light reaching us from a star. In fact, the life story of a star is written in its light. Most of those special instruments did not exist even a hundred years ago. And we have had telescopes for only about 400 years. Powerful telescopes have been around for less than a hundred years. The first space telescope—the Hubble Space Telescope—was put into orbit in 1990.

Ziggurats and Kings

Without telescopes or other instruments of any kind, stargazers of 3,000 and 5,000 years ago could only wonder about the stars. Even so, we can be sure they were expert observers because they relied on the stars to find their way across deserts when travel was comfortable only during the cool of night. And later, they used the stars for navigation at sea. But long before the days of sailing ships the stars were regarded in a very different and even more important way. They were thought to be abodes of the gods.

If we turn the pages of the calendar back to about 7,000 years ago, we find people called the Sumerians, whose writing on clay tablets is the earliest writing we know of and who left records of what they saw in the night sky. Where these people came from is a mystery. About all we can say is that by at least 5000 B.C. they were settled in the Middle East between the eastern end of the Mediterranean Sea and the Persian Gulf, in what is now Iraq. The Sumerians were skilled farmers at a time when many other peoples still lived off the land as nomad hunters who followed the game and gathered wild food. The Sumerians looked down on such wandering tribes, and their written records refer to them as "people who do not know houses and who do not cultivate wheat."

People such as the Sumerians who depended on planned agriculture had to be expert sky watchers. They relied on their knowledge of the motions and changing positions of the constellations, Sun, and Moon to signal when to plant their crops and so start the agricultural season, and when to harvest. "Season" is the key word here. It was essential that they knew the safest time to plant so that their seeds would survive and provide an abundant harvest. The job of drawing up the first calendars to keep track of the changing seasons fell to the astronomer-priests. Because of their detailed knowledge of the sky, they were men of great power, both in the eyes of the common people and especially in the eyes of their kings and emperors. The calendars were used not only to signal the best time to plant but also to mark the occurrence of unusual sky events such as eclipses and comets.

The Sumerians eventually were conquered by neighboring

people known as the Babylonians. They adopted many ways of the Sumerians, including certain aspects of their religion. The Babylonian astronomer-priests made their observations from great towers, called *ziggurats*, built out on the sprawling flats where they had a clear view of the sky in all directions. Each time they made their observations of the heavens they noted any and

Some 5,000 and more years ago people in the Middle East built towering structures that were used as observatories to view the stars and planets. Ancient Babylonian astronomer-priests drew up early calendars based on the changing phases of the Moon. Called ziggurats, the towers also were places where the priests supposedly met and communicated with the gods.

all earthly events then taking place—change of seasons, earth-quakes, floods, epidemics, victory or defeat of armies, and the death and birth of kings. Their chief reason for observing the sky, however, was to keep track of the passage of time in order to accurately fix the dates of religious festivals, during which the holy men communicated with the gods. By about 1700 B.C., the Babylonian astronomer-priests had recorded more than 7,000 celestial observations and omens collectively called the "Enuma Anu Enlil."

Early on, the Babylonians regarded the Sun, Moon, and planets as spiritual dwellings of the gods. Later, however, they came to regard each sky object as a material object ruled by a particular god rather than as a god itself. The gods, then, were made into spiritual beings without substance. The Babylonian astronomer-priests believed that the same divine powers that ruled events in the heavens also ruled events on Earth. They further believed that the gods met once a year to decide the fates of nations and kings. The meetings took place in the ziggurat obser-vation towers, which were holy mounds up to 300 feet (91 meters) high. It was said that atop the ziggurats holy men were able to communicate with the gods.

Shang di and the Son of Heaven

This notion of holy men communicating with the gods also is found in ancient Chinese astronomy. As with the Babylonians, the ancient Chinese keepers of the calendar were men of power. Successful agriculture for them, too, depended on a detailed and accurate calendar to mark when the seasons changed. Each Chinese emperor had a court of several astronomers. Some were specialists on the importance of the changing night-to-night posi-tion of a single planet. Others concentrated on eclipses or comets.

By 400 B.C. Chinese comet watchers had made drawings of 29 different kinds of comets, many warning of calamity. Still other astronomers compiled omens—forewarnings of defeat or success in battle. The outcome of a planned battle was supposed to depend on the planets' positions among the stars at the time of battle. That superstitious belief, called *astrology*, was based on the belief in spiritual powers associated with the planets and stars.

The practice of astronomy, which recorded the changing positions of the constellations and planets from one season to the next, was very different from astrology. But the two were hopelessly mixed up and served as a double-edged sword. Since the sky was looked on as holding power over the fate of the state, and of the emperor, those who could "correctly" read and interpret sky events were much favored by the emperor. Those who misread the sky usually lost their lives.

Even though each emperor was regarded as godlike, he was subject to the gods' displeasures if he acted foolishly or ruled unwisely. Because the interpretations of sky events by the court astronomers could also spell the success or failure of an emperor, the astronomers often held the fate of the emperor, or the very state, in their power.

Each emperor was thought to have magical powers that enabled him to interact with the sky spirits. As the sky gods ruled from above, the emperor—called the "Son of Heaven"—was their flesh and blood counterpart who ruled below.

That arrangement could be seen in the sky itself. Like ancient astronomers in other parts of the world, the Chinese regarded Earth as the center of the Universe. After all, anyone could see that all objects in the heavens circled around a motionless Earth at the center of creation. The Chinese called their land the Middle Kingdom, reflecting their belief that China marked the

There are many Native American creation myths. In one Tlingit story, a raven steals the Sun, Moon and stars from a box where they had been hidden by a powerful chief who wanted to keep the world in darkness. During the raven's escape flight from the angry chief, the bird scattered his celestial beacons and so brought light to the world.

center of all creation. And China's capital, Beijing (known then as Peking), was located in the center of that greater center. Further, the emperor, surrounded by his court of astronomers, was the ultimate central point in his Imperial Palace, called the Forbidden City.

That neat and tidy earthly arrangement of authority had its larger arrangement in the sky. Polaris, presently the North Star, appeared to the ancient Chinese hardly to move in relation to the other stars. It was supposed to be the radiant face of Shang di, the God of Heaven. The emperor was his earthly counterpart, the Son of Heaven. The Forbidden City, with its court of astronomers surrounding the emperor, was the earthly counterpart of the stars that forever appeared to circle Polaris. Today we call that star group the north circumpolar (meaning "around the pole") stars. To the Chinese, that region of circumpolar stars clustered about the motionless North Star heavenly hub was the Polar Forbidden City. The American astronomer E.C. Krupp describes this and other aspects of Chinese and Babylonian worldviews in fascinating detail in his book, *Skywatchers, Shamans & Kings*.

In still earlier times, the Chinese thought of the stars as souls waiting to be born. In the ancient Chinese *Annals of the Bamboo Books* are accounts of the birth of a line of emperors. All of them supposedly were souls in the form of fiery light, or shooting stars. At the right moment, a shooting star revealed itself and entered the human embryo being carried by the mother of the next emperor-to-be. The emperors so born entered this world at various times from 2698 B.C. to 2205 B.C.

Before leaving those ancient Chinese astronomers and the astronomer-priests of the Middle East, we should emphasize one important thing. Their interest in astronomy had nothing to do with trying to find out what the stars were made of or their actual arrangement in space. Their interest was focused on how the heavens were supposed to direct affairs on Earth. The expert sky watchers of those ancient times felt that the more accurate their observations and interpretations, the better they were serving their kings and emperors spiritually. An example of how this idea led to highly accurate observations is that Chinese calendar makers of the year A.D. 1199 measured the length of the year as

Circles in the sky are made when a camera pointed at the Polaris, the North Star, takes a long time exposure photograph. All other stars appear to circle Polaris (the faint star in the center of the circles just above the large tree). Circular star trails are traced out as Earth rotates on its axis; so it is Earth's motion, not the motion of the stars that makes the trails. The color of each circular trail reveals the color of the star making the trail.

365.2425 days. That is only *26 seconds* off today's value, and it is exactly the value we use in our modern calendar!

The journey from atop the ziggurats of the astronomer-priests of Babylon, and from the Polar Forbidden City of the Chinese, to the schools of the early Greek philosophers was long and difficult. At its end, the old sky gods thought to govern affairs on Earth were toppled from their celestial perches, and astronomy was put on a somewhat firmer foundation.

Two

The Greeks, Romans, and Arabs

The period of ancient Greek astronomy began with the Golden Age of Greece between 600 and 500 B.C. The great thinkers of these ancient times were not scientists, as we think of scientists today. They were philosophers who tried to understand some grand plan of the world and everything in it. There were no sciences of chemistry, physics, biology, or meteorology. Their only tools were mathematics, their keenness as observers, and their imagination.

Moistures and Fire

Among those early thinkers was a teacher named Anaximenes who lived around 540 B.C. He pictured planet Earth as a flat tabletop with moisture rising from it here and there. The moisture, he said, formed the stars. And since the moisture-substance rose, there could not be any stars beneath the Earth-table. The stars, he explained, move around Earth "as a cap turns round our head." He explained the Sun's disappearance at night by saying that it "is hidden from sight, not because it goes under Earth, but because it is hidden by mountains, and because its distance from us becomes greater then."

Heraclitus, another Greek scholar who was born about 540 B.C., said that the stars, including the Sun, were made of fire resting in a bowl. At night when we can see the stars, he explained, the mouths of the bowls are turned toward us and we see the stars' fire. But during the day, the bowls turn upward and hide the fire from view. Because the Sun-bowl was closest to Earth, he said, we received more light and heat from it than we do from the more distant bowls. Here was very early evidence that the Sun was thought of as being a star.

Empedocles, who was born about 490 B.C., pictured the Universe as a great, slowly-turning shell, half of which was lighted inside by fire. Located at the center of the shell was Earth. It was daylight whenever the fiery half of the shell was above Earth. But as the shell gradually turned, the fiery half slowly disappeared beneath Earth and was replaced by the dark half. The Sun, he said, was not an object at all, but only a reflection of the fire source by the sky-shell.

Some thought that the stars were glowing balls of molten iron, an idea that was to last many centuries. Actually, the idea isn't so farfetched, especially if you have ever seen a meteor shower. When iron meteorites strike the ground, they are hot. They are hot not because they were hot stars, but because their swift trip down through Earth's atmosphere heated them. To some, it seemed perfectly reasonable to think that the meteors we commonly see streaking their way down through Earth's air were "falling stars."

"Shooting stars" is the old name for meteors. Until about the 1800s, when scientists came to know the true nature of meteorites and meteors, people commonly thought that meteors actually were stars falling from the sky. During one especially rich meteor shower some people thought that the next night there would be no stars left in the sky.

The Great Aristotle

Aristotle is regarded as the most famous philosopher the world has produced. Born in 384 B.C., he was a likable youth, pleasant and both quick and eager to learn. He was the favorite student of the great philosopher Plato. When Plato died in 347 B.C., Aristotle left Plato's Academy in Athens and struck off on his own.

Aristotle was a man of logic and reason and looked down on the superstitious beliefs of the uneducated. He most likely scorned the old popular belief that important people turned into stars when they died. Earlier, the Greek playwright Aristophanes, in his comic play, *Peace*, assured his audience that servants as well as their masters and national heroes all may look forward to stardom. At least a thousand years earlier, the Egyptians were writing accounts of the dead being "reborn" as stars. The Egyptian hieroglyph from which we get our star sign ✶, means "soul."

Aristotle accepted an earlier idea that all earthly matter was made of combinations of four basic "elements"—earth, air, fire, and water. For instance, paper was said to be part fire and part earth because when it was burned flame escaped from it, and the ash that was left was earth. The true chemical elements were not to be identified for hundreds of years later. Aristotle taught that everything had its own "natural place." Since heavy matter such as metals sank, it was locked up in the rocks beneath Earth's surface. Lighter air-matter floated about Earth's surface. Fire was still lighter and could be seen by anyone to rise into, and presumably above, the air where it formed a sphere of its own.

Aristotle accepted the idea that Earth did not move in any way and that it formed the center of the Universe. The Moon, Sun, planets, and stars—in that order—all revolved about Earth. All matter this side of the Moon, he said, was impure and capable of being changed. Such matter included comets, rainbows, meteors, and lightning. He believed that comets were warnings from the gods. The motion of all such impure matter, he taught, was not regular and could follow paths of different shapes—straight lines, curves, zigzag patterns, and so on. But beyond the Moon, in the heavens, things were very different. All heavenly matter was pure, never changed, and moved steadily in circles. Aristotle looked on the circle as the most perfect of all forms, for it had no beginning and no end. This went nicely with his notion that the Universe was older than anyone could possibly imagine and goes on possibly without beginning or without end.

Matter making up the stars and planets, he thought, must be of a kind very different from the four impure base elements. He invented a fifth and the purest of pure elements called *quinta essentia*, meaning "fifth essence." Why did the stars shine? It was their motion across the sky that caused them to give off heat and light, he explained. They rose in the east and set in the west because they were attached to a huge sphere that at the beginning of creation was set turning by some divine power. What that power was he does not say. And the stars twinkle because of their very much greater distance than the planets. When we look across that great distance, he explained, starlight "wavers on account of its weakness." Today we know it is the motion of the atmosphere that makes the stars seem to twinkle.

Aristotle reasoned that since all the stars are attached to a huge wheeling sphere, then the Universe itself must have the shape of an even bigger sphere. He never could accept the idea that the Universe was either infinitely large or infinitely old.

Although there were a number of Greek astronomers who lived after Aristotle's time, none showed any particular interest in what the stars were made of, what made them shine, or their distances. It is important to remember that at the time of the ancient

Greeks there was no science of chemistry or of physics. Without those two sciences, that branch of astronomy called astrophysics couldn't get very far. The Greeks were excellent mathematicians, and mathematics was the only foundation stone they could give astronomy at that time. So it is not surprising to find them at least trying to measure the distance to the Moon and Sun.

One of the greatest astronomers of ancient times was Hipparchus, who lived around 150 B.C. His method of measuring the Sun's distance from Earth was to time the Moon's passage through Earth's shadow during a lunar eclipse. Even though the idea was sound, the measurement was too hard to make accurately. His distance to the Sun came out to only 9 million miles (15 million kilometers). Although 10 times too short, it was the best measurement at hand, and it provided a start. Further, it suggested that the Universe was very much larger than anyone had suspected.

After the death of Hipparchus in 127 B.C., the practice of astronomy in Greece came to an end as the Romans began to gain political control. They were more interested in literature, politics, and history than in science. Since mathematics was not one of their strengths, observational astronomy fell by the wayside. In its place came a rebirth of astrology and with it a return to the stars and planets as spirits that influenced our lives.

Stars: The Abodes of Souls

The Romans' chief interest in the stars was astrological, not astronomical. If you had a question, any question, just ask the stars. Or more properly, ask your astrologer who claimed to know how to interpret what the planets and stars would be up to when you planned a trip, or set a date for your wedding, or placed a bet on the chariot races.

The Romans delighted in the old belief that departed souls

became stars, but not just any old soul. You had to be an important person to gain stardom. Comets and "shooting stars," which are meteors, were thought to be the souls of heroes on their way to heaven or returning to Earth to take up life again. In 44 B.C. a comet observed after the Roman emperor Julius Caesar was murdered was named *Sidus Julium* ("Caesar's Star"). Supposedly it was Caesar on his way to heaven. Around the same time, the Druids, who lived in Britain, worshiped the stars as abodes of departed souls. A person's soul, they believed, moved from one star to another, each move raising it to a grander state of spiritual being.

Because the stars were unknowable as physical objects to people up to the time of the Romans, and much later, the stars served as an excellent means of wish fulfillment for an afterlife. People don't like sad endings, and the elusive stars suggested a future of some sort. In the case of the "hairy stars," which comets were called, an actual rebirth in human form was promised.

About a hundred years after Caesar's Star carried the dead emperor away, to where we cannot be sure, the last of the great astronomers of ancient times lived. His name was Claudius Ptolemaeus, and the time we assign to him is A.D. 150. In a huge book called *Almagest*, Ptolemy summarized all of ancient Greek astronomy. He praised Aristotle and put his own stamp of approval on a Universe that had motionless Earth at the center. Ptolemy was more interested in improving the way to predict a planet's motion than in what the stars were made of, what made them shine, or their actual locations in space. Like Aristotle and the other greats of ancient Greece, Ptolemy simply accepted the idea that the stars were attached to a huge crystalline sphere that turned and so carried them across the sky from east to west. After Ptolemy's death there was virtually no new work done in astronomy in the Western world for a long time. As the late historian of astronomy Colin A. Ronan has put it, "On Ptolemy's death the light of astronomical research went out for a thousand years."

The "House of Wisdom"

From the time of Ptolemy's death up to about the 1500s, Christian and Islamic religious leaders preached that the only worthwhile knowledge was the knowledge of God. If the books

Ptolemy's plan for a universe in which Earth stood still at the center with the Sun and planets circling our planet became the official and approved view of the church until about 1600. In this illustration from Martin Luther's Bible (1534), God looks down on his handiwork. Adam and Eve stand in their Eden paradise surrounded by animals and creatures of the sky. The Moon, Sun, and stars are all shown to circle a motionless Earth.

of earlier times did not praise God's work, then the books must be burned. So one after another, the great libraries and centers of learning were burned to the ground, including the greatest one of all, in Alexandria, Egypt. In those days books could not simply be replaced. The printing press had not been invented. There was no way to make hundreds of copies of a book overnight. Scribes had to hand copy each manuscript, which took days or weeks depending on how long the manuscript was.

In the early 800s the enlightened Arab ruler, Caliph Haroun-al-Rashid, did not approve of the burning and destruction. He provided funds for his son, the Caliph Al-Mamun, to start a library in Baghdad, called the House of Wisdom. Scholars came from far and wide and brought priceless copies of old Greek

During the Dark Ages, around A.D. 1000, Christian and Islamic religious leaders ordered the destruction of libraries and all scientific books that did not praise God. Some enlightened Arab leaders, however, saved many of the ancient texts and had them translated into Arabic. This Arabic illustration from the 1500s shows astronomers at work in the observatory in Istanbul. Four hold instruments used to measure the height of stars and planets above the horizon.

books. They were translated into Arabic and so preserved. Among them was Ptolemy's *Almagest*, actually given that title by the Arabs when they translated it from Greek into their language. In Arabic, the title means "The Greatest."

The Arabs made many fine instruments to measure star and planet positions, but they showed no interest in learning anything about the stars themselves. One of their chief concerns was to improve the accuracy of their Moon calendar. They had long used a Moon calendar to mark the times for their religious ceremonies. The same had been true of the Babylonians some 3,000 years earlier. So, the Arabs brought astronomy out of cold storage and slowly revived it. To this day, stars with Arabic names—such as Algol, Zubenelgenubi, Betelgeuse, and others—remind us of the role the Arabs played in preserving the astronomical knowledge of the ancient Greeks up to the time of Ptolemy.

Before leaving the Arabs, we should mention another of their contributions, even though it has nothing to do with stars, directly. They introduced, from India, the use of what we now call Arabic numerals, the numbers we use today in mathematics. They made it a simple matter to add, for example, a series of numbers such as 7, 78, and 779. Doing that task with Roman numerals was far more cumbersome. Try adding VII, LXXVIII, and DCCLXXIX. And it was the Arabs who developed the mathematics of algebra, the word coming from the Arabic words *al-jebr*, meaning "bone setting, or reunion of broken parts."

New Eyes on the Stars

We now jump ahead to the 1600s when two things happened to kindle a new interest in the stars. One happened gradually, the other almost overnight.

In the early 1600s a Dutch maker of eyeglasses discovered

that by holding two lenses one in front of the other, distant objects appeared closer. Around 1609 the Italian astronomer Galileo Galilei heard of the discovery, made a telescope, and became the first to publish what this marvelous new instrument revealed about the heavens.

Galileo had the habit of getting into trouble by rubbing his fellow university teachers the wrong way. He took great pleasure in poking holes in some of the ideas of Aristotle, whose authority was still very much respected during Galileo's time. For instance, his telescope showed that the Sun and Moon were not the unblemished objects of perfection that Aristotle had thought them to be. The Sun had dark spots that came and went and that were carried around the Sun once about every 27 days. The Moon had thousands of scars in the form of craters, deep cracks in its surface, and mountains like those on Earth. And when Galileo pointed his "optik tube" at that hazy band of light called the Milky Way, he was astonished to find that it was made up of

Galileo became the first to study the night sky with a telescope, around 1610. His drawings of the Moon at various phases detailed craters and sprawling dark misnamed "seas." These drawings appeared in his book The Starry Messenger, *published in 1616.*

countless individual stars. Galileo didn't help his cause by pooh-poohing Aristotle's and Ptolemy's claim that a motionless Earth sat at the center of the Universe. Although Galileo couldn't prove it, he taught that the Sun was at the center and that Earth moved around the Sun. He was, however, able to show that Venus circled the Sun, not Earth.

Galileo taught that the old view of Aristotle and Ptolemy that a motionless Earth stood at the center of the universe was wrong. Instead he championed the view of the Polish astronomer Nicolaus Copernicus (1543) who said that the Sun marked the center and that Earth, all the other known planets, and the stars circled the central Sun. This Dutch engraving done in the 1600s correctly shows the order of the planets. It also shows Earth at its two equinox positions (spring and fall) and two solstice positions (winter and summer). Notice also the circular band of the constellations of the Zodiac wrapped around the Solar System.

Around Galileo's time a brilliant young mathematics teacher named Johannes Kepler shattered another of Aristotle's ideas— that the planets move in perfect circles. A firm believer in a Sun-centered system, Kepler was able to show that the planets move in paths called ellipses, which are not perfect circles. Certain other scholars also began to favor the idea that Earth revolved around the Sun and that it turned on its axis like a spinning top. It was that spinning motion that made the stars only

*Although Galileo could not prove that Earth circled the Sun, rather than the Sun and other planets circling Earth, he was able to show that Venus circled the Sun. Careful observation of Venus showed that it went through phases, which meant that it had to be circling the Sun, not Earth. The easiest phase for Galileo to see was Venus's gibbous phases, shown at positions **A** and **B**.*

seem to parade across the sky as a group. Since that was so, then there was no longer a need to have the stars attached to a great spinning crystalline sphere. So that ancient notion also came crashing down, and with it the idea that the stars were all the same distance from Earth. Freed from their crystalline sphere merry-go-round, the stars could lie at many different distances. Furthermore, the Universe no longer needed to be limited in size. Space could be infinite. (Aristotle would have been troubled.)

Those ideas fired the imaginations of astronomers and philosophers all over Europe. One such thinker was the French philosopher and mathematician René Descartes, who lived from 1596 to 1650. Although his ideas about stars sound strange to us today, his thinking was an attempt to solve an age-old puzzle. Space, he said, was cluttered with moving matter that here and there collected and formed "whirlpools." At the center of each whirlpool a star formed. Gradually a kind of skin grew over each star. Evidence for the beginning of such a process, he said, were the spots Galileo had seen on the Sun. Eventually a star stopped shining and its whirlpool collapsed when the skin completely encloses a star. The "dead" star then either moved to another whirlpool as a comet, or it settled down in an orbit as a planet. He believed that moons of the planets, including Earth's Moon, were old worn-out planets. No one since the time of Aristotle had been bold enough to put together a grand scheme that included all matter in the Universe.

Since the Sun was the nearest star, it is not surprising that scientists after Descartes' time turned their attention to the Sun: What was it made of? What made it give off heat, light, and other energy? How long had it been shining? And how much longer could it go on shining?

Three

A New Look at the Sun

For many centuries people thought that the Sun, and other stars, might be great balls of fire. What else could produce the enormous amounts of energy given off by the Sun? Today we know that in one second the Sun gives off more energy than people have used ever since there have been people on the planet. In one heartbeat the Sun pours out as much energy as the explosion of 100 million hydrogen bombs. To the ancient Greeks, star fire was a very special substance unlike anything else on Earth, and certainly not like the fire from a burning candle or log.

The Sun Is Too Hot to Burn

If the Sun gives off energy by burning, what could be burning? A great pile of wood? An enormous lump of cosmic coal? Scientists

This drawing of the Sun made in the year 1635 suggests that it was a fiery object with smoke-like puffs being given off. Mountains and other Earthlike features were imagined to decorate the Sun's surface. The idea of the Sun as a burning object remained popular for many years.

of the 1700s and 1800s had been asking that question. Whatever was burning was packed into an enormous ball that had been burning longer than anyone could remember. And it had been burning rather steadily.

Records going back some 5,000 years do not mention a Sun that was either very much brighter or very much dimmer that the one we see today. But we have records going back even longer than that. Biologists know of fossils of a marine organism called *Neopilina*, which lived 500 million years ago. *Neopilina* organisms continue to live today unchanged from their ancestral forms. If the Sun had been very much brighter or dimmer than now, those organisms probably would not have survived the change. So, at least half a billion years ago the Sun was just about the same as it is now. If it were a great pile of wood or coal we would

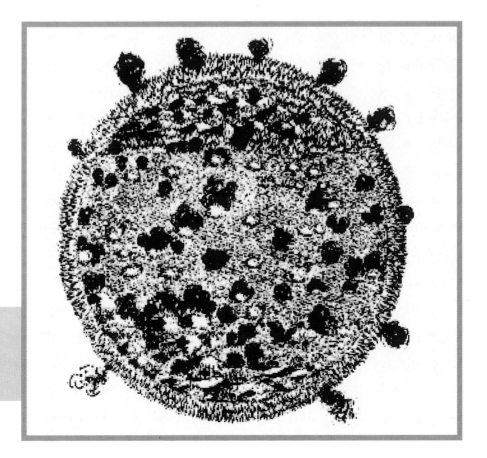

expect the fire to dim over that very long period of time. In fact, we can compute how long a burning Sun giving off the amount of energy that it does could last. It turns out to be only a few thousand years.

Here's another convincing reason why the Sun cannot be burning. We know that burning takes place when whatever burns combines with oxygen in the air, called *rapid oxidation*. Atoms of oxygen and wood, for instance, combine at fairly low temperatures below about 5,370°F (3,000°C). At temperatures much higher than that atoms cannot combine. Instead, their rapid motion and forceful collisions cause them to break apart. Oxidation is no longer possible. Since the Sun's surface gases are about 10,000°F (5,600°C), the Sun is too hot to burn.

The idea that the Sun was an enormous chunk of a burning substance fell out of favor in the early 1800s. Scientists at that time calculated that a burning Sun could last only a few thousand years and would gradually grow dimmer and cooler. Today we know that the Sun actually is some 30 percent brighter than it was several billion years ago. Earth's atmosphere over that long period of time has been responsible for regulating the amount of energy reaching the planet's surface.

How else might the Sun be giving off its tremendous energy, of which Earth—fortunately—receives only the tiniest fraction. In 1848, the German physician Julius R. von Mayer guessed that millions of meteorites regularly crashed into the Sun and so kept it hot and made it shine. But he had to give up the idea when he couldn't explain why Mercury, Venus, and Earth had long been targeted by meteorites but were not also heated enough to shine.

Could the Sun Shine by Shrinking?

In the 1850s, the British scientist William Thomson (Lord Kelvin) and the German Hermann von Helmholtz tried their hand at solving the solar energy problem. Chemists had long known about gases and how they behaved. The properties of oxygen and hydrogen were well known, and the role played by oxygen in the process of burning was well understood. The French chemist Antoine Lavoisier had seen to that around 1765.

So Thomson and Helmholtz had good grounds for correctly assuming that the Sun was a great ball of hot gases. Is it possible, they wondered, that the Sun is slowly contracting under its own gravity? That would cause an enormous buildup of pressure deep in the Sun. Was it then possible that the Sun gave off all its energy as a result of its gases being squeezed tighter and tighter around its core? A contracting Sun would produce about 10,000 times more energy than a burning Sun. The idea seemed convincing and was popular for many years. But was it right?

At the time Thomson and Helmholtz lived, scientists had a pretty good idea of how much energy the Sun puts out, based on Helmholtz's contraction theory. It even provided an insight into how stars evolve. The contraction model seemed so promising that it became the standard model until almost 1920. In that year the British astronomer Sir Arthur Eddington delivered a truly remarkable paper. In it he said, "Only the inertia of tradition keeps the contraction hypothesis alive—or rather, not alive, but an unburied corpse."

If the Sun and other stars gave off energy by contraction, then they must gradually be shrinking year by year. To pour out as much energy as the Sun does, it would have to shrink by about 66 feet (20 meters) in diameter each year. Although that's not very much in one year, it adds up over the hundreds of millions of years the Sun has been shining at its present rate of energy output. It turns out that at that rate of energy production the Sun would run out of diameter, or be completely used up, in only a hundred million years. The fact that biologists knew of fossils more than 30 times older than that made the idea of a contracting Sun impossible. Another problem was that to keep up its energy production as we see it burning today, only about 20 million years ago the Sun would have had to have been big enough to fill Earth's present orbit!

It wasn't until later in the 1900s that someone came up with the fine details of how stars shine, although Eddington had worked out the essential idea—a new source of energy far greater than gravity. And in the process he even predicted the hydrogen bomb! Before finding out what that idea was, we should first look into some discoveries that made *that* discovery possible.

There's More to Light Than Meets the Eye

Imagine Isaac Newton's surprise when he discovered that light could be decomposed. *Decomposed?* At the time Newton experimented with light, in 1665, no one knew what light was. Today we know a lot about light. For example, we know that light behaves as if it were made up of particles like tiny BB pellets. But at the same time it behaves like a train of moving ocean waves with crests and troughs. We also think that nothing in the Universe can travel faster than light—186,282 miles a second (299,792 kilometers a second). Newton did not know either of those things about light.

However, by shining light through a prism he showed that white light is made up of a set of rainbow colors called the *visible spectrum*, with red at one end and violet at the other. While all of those colors are part of sunshine and starshine and travel at the

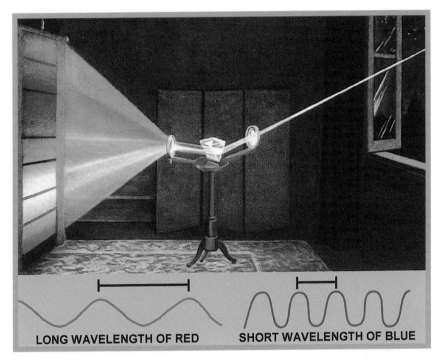

LONG WAVELENGTH OF RED   **SHORT WAVELENGTH OF BLUE**

In 1665 Isaac Newton discovered that the white light from the Sun, or from a distant star, could be separated into a rainbow of colors called a spectrum. Later scientists found that each color could be measured in units called wavelengths, similar to the lengths between the crests of ocean waves. At the red end of the spectrum visible to the eye the wavelengths were longest. They then became increasingly shorter toward the blue and violet end of the visible spectrum. Wavelength is measured from one wave crest to the next crest.

speed of light, they have an important difference, called *wave-length*. Wavelength is simply the distance from the top of one wave crest to the top of the next wave crest. That is true in a wave train of red light, or yellow light, or blue light, for example. In the visible spectrum red light has the longest wavelength and is the least energetic. At the opposite end is the much more energetic violet light with the shortest wavelength.

Beyond both ends of the visible spectrum is energy of longer or shorter wavelengths. That larger energy spectrum is called the electromagnetic spectrum. Beyond violet light is invisible energy of shorter wavelengths called ultraviolet, Xray, and gamma ray energy. At the opposite end beyond red light are longer wavelengths called infrared, or heat, energy. Beyond that are the wavelengths of microwave energy and then radio waves. Energy all along the electromagnetic spectrum is the total energy that reaches us from the Sun and other stars.

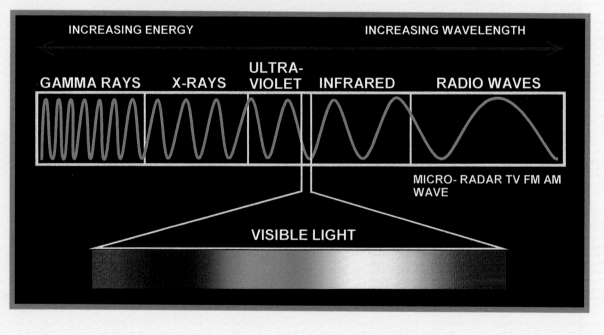

INCREASING ENERGY   INCREASING WAVELENGTH

GAMMA RAYS   X-RAYS   ULTRA-VIOLET   INFRARED   RADIO WAVES

MICRO- RADAR TV FM AM
WAVE

VISIBLE LIGHT

Although Newton knew that energy in the visible spectrum came from the Sun, he did not know that the visible spectrum is only a small band of energy within a much larger spectrum of star energy. That larger spectrum is called the *electromagnetic spectrum*. Beyond violet light is invisible energy of still shorter wavelength called ultraviolet. That is the light, or radiation, that gives you sunburn. Beyond the ultraviolet are the still shorter wavelengths of Xrays, and at the high-energy end gamma rays. Gamma rays have the shortest wavelengths of all and are the most destructive.

In the opposite end beyond red light are longer wavelengths of energy called infrared, which is heat energy. At still longer wavelengths are microwaves, which heat a bowl of soup in your microwave oven. And at still longer wavelengths are radio waves that enter your FM tuner and come out as sound. Energy all along the electromagnetic spectrum is the total energy that reaches us from the Sun and all the other stars. Since Newton's time, astronomers have studied the messages of starlight to learn about what the stars are made of and what makes them shine.

Fingerprinting the Stars

Newton and those after him in the 1700s saw only a smooth spectrum of rainbow colors, called a *continuous spectrum*. It was continuous simply because it was not broken up in any way.

But then in the 1800s an instrument called a *spectroscope* was invented. When attached to a telescope aimed at the Sun or some other star, a spectroscope produced a spectrum broken up by a pattern of lines. The line pattern was always the same for one chemical element but different for a different chemical element. Hydrogen, for instance, showed one pattern of lines, but the chemical element sodium showed a very different line pattern. By the

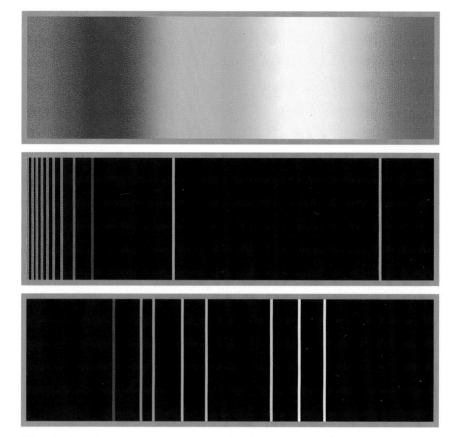

White light passing through a prism is displayed as a rainbow of colors called a continuous spectrum (top). When the Sun is examined through an instrument called a spectroscope, the device shows the "fingerprint" patterns of chemical elements in the Sun. Each element has its own individual pattern. The pattern for hydrogen is shown at center, and that for sodium at bottom. (The bright line patterns for hydrogen and sodium are approximate, not exact.)

mid-1800s astronomers had learned that each chemical element had its own private set of "fingerprints" in the form of spectral lines. For the first time, it was possible to point a telescope at a star and tell what kinds of gases the star contained. More than 70 of the more than 100 known chemical elements have been found in the Sun and other stars. Most of those gaseous elements are the lightest chemical element hydrogen (70 percent) and the second lightest, helium (26 percent).

The spectral line patterns of hundreds of stars were photographed in the late 1800s, and thousands more in the early 1900s. There were so many different line patterns that it soon became a major task to decide how to group the stars as belonging to one class or another class. In 1891 the American astronomer Edward C. Pickering began to organize stars into classes based on the strength of their hydrogen lines. Stars such as Sirius and Vega, with their especially strong hydrogen lines, became Class A stars. At the other end of his list came stars like Betelgeuse with their very weak hydrogen lines. They became Class M stars. And there were other classes: B, C, D, E, F, and so on up to Class Q stars, which had few or no hydrogen lines.

Pickering's assistants included three female astronomers named Williamina P. Fleming, Antonia Maury, and Annie Jump Cannon. The team felt the original system was too cumbersome and simplified it to include only seven classes arranged in the seven spectral line patterns called OBAFGKM. As the table shows (p.43), O stars are the hottest and shine with a blue light. B stars are the next hottest and shine with a bluish-white light,

Professor Edward C. Pickering was director of the Harvard Observatory. He led a team of researchers who grouped stars into various classes based on the strength of a star's hydrogen lines in its spectrum. Pickering lived from 1846 to 1919.

During her lifetime, Dr. Annie Jump Cannon became one of the world's
most highly respected astronomers. A member of Pickering's team and
curator of the Harvard Observatory Record, she classified more than
350,000 stars. She lived from 1863 to 1941.

and so on to M stars, which are the coolest and shine with an orange-red light. By the time she died in 1941, Annie Cannon had classified more than 350,000 stars! She is one of the most highly respected astronomers in history. (Hipparchus would have been astonished.)

One important thing that astronomers learned by studying the spectral line patterns of stars was that a star's color tells us the temperature of its surface gases. The temperature of its surface gases then became a clue to what must be happening deep inside the star. If the surface gases are very hot, then the temperature

Star Class	Color	Surface Temperature
O	Blue	28,000 to 50,000 kelvins*
B	Blue-white	9,900 to 28,000
A	White	7,400 to 9,900
F	Yellow-white	6,000 to 7,400
G	Yellow	4,900 to 6,000
K	Orange	3,500 to 4,900
M	Orange-red	2,000 to 3,500
R,N,S	Red	2,000 to 3,500

Temperatures are given in "kelvins," or degrees on the absolute temperature scale. On this scale water boils at 373°. Under Star Class, star types R, N, and S are subclasses of M-type stars.

inside the star must be still hotter. That is because heat flows in only one direction—from a region of high temperature to a region of lower temperature. That's how a furnace keeps your house warm. High heat deep within a star could be caused by only one thing. The gases making up a star had to be pressing down into the star with tremendous force and packing all that star matter in the core region with an even greater force.

With that general view of life inside a star, clues to what makes a star shine began to emerge in the early 1900s.

Meet Mr. Einstein

In 1905 one of the greatest physicists of all time, Albert Einstein, wrote a simple equation that paved the way for our understanding of what makes stars shine. The equation said matter could be changed into energy and that energy could be changed back into matter. Further, that huge amounts of energy could come from an extremely small amount of matter. For example, when changed

In 1905 Albert Einstein paved the way for an understanding of how stars shine. He showed how matter could be changed into energy and that energy could be changed back into matter. A very small amount of matter, he said, could be changed into an unbelievably large amount of energy. His work led to an understanding of how the Sun maintains life on Earth.

completely into energy, a small stone could supply all the electrical needs of an average household for a million years.

Fifteen years later in England, the British physicist Sir Arthur Eddington worked out a problem that might be called "The Atom with the Missing Mass." He knew that 4 hydrogen atoms weighed a certain amount. He also knew that one helium atom could be built up from 4 hydrogen atoms. BUT, and here was the mystery, the resulting helium atom weighed less than the 4 hydrogen atoms that made the helium atom. It weighed 0.7 percent less. Why? His answer was that somehow during the conversion of the hydrogen atoms into helium, the 0.7 percent of missing mass was changed into energy. And that was just what Einstein had said was possible.

By the late 1920s Cecilia Payne-Gaposchkin, of Harvard University, showed that stars are made mostly of hydrogen that seemed to be changed into helium by the fusion of hydrogen atoms. If all of the Sun's hydrogen took part in fusion reactions, then the Sun could shine for some 100 billion years. But Payne-Gaposchkin showed that only about 0.7 percent of the Sun's store of hydrogen in the core might be available for hydrogen fusion into helium. *That would release enough energy to keep the Sun shining for only 10 billion years.* Could the fusion of hydrogen into helium possibly be what made the Sun and all the other stars shine? (Aristotle would have been impressed. Eddington knew it all along.)

The scene next moves ahead to the early 1930s. Three main players answered yes to that question and explained just how the

The four simplest of well over a hundred different kinds of chemical elements are hydrogen, helium, lithium, and beryllium. Hydrogen, the simplest atom of all, has a single electron held at a certain energy level distance from a single proton. Notice that in all four atoms each has the same number of electrons as protons. Because the negative charges of the electrons cancel out the positive charges of the protons, each atoms is electrically neutral. Neutrons lack any electric charge and so simply add mass to an atom.

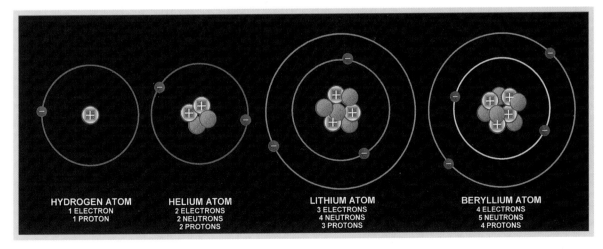

HYDROGEN ATOM
1 ELECTRON
1 PROTON

HELIUM ATOM
2 ELECTRONS
2 NEUTRONS
2 PROTONS

LITHIUM ATOM
3 ELECTRONS
4 NEUTRONS
3 PROTONS

BERYLLIUM ATOM
4 ELECTRONS
5 NEUTRONS
4 PROTONS

Sun and other stars pour forth such enormous amounts of energy. They were the American astronomers Hans Bethe and George Gamow, and the German Karl von Weizsacker. Their explanation was a series of atomic reactions called the *proton-proton chain*. It was an idea that seemed to work well, the idea that Eddington had proposed more than a decade earlier, and which just about all astronomers accept today.

The Proton-Proton Chain

The Sun's energy furnace is in its dense core region, they said, and it is fueled by some of the Sun's vast store of hydrogen. If you don't know how a hydrogen atom is put together, you are about to find out. The hydrogen atom has one lump of matter called a *proton* forming the center, or *nucleus*, of the atom. It also has a much smaller and lighter particle called an *electron*. The electron hovers in a shell surrounding the nucleus. At room temperature the electron and proton are held close to each other by their opposite electrical charges. The proton has a plus charge while the electron has a minus charge. But inside the Sun things are not at room temperature and are different enough to upset this tidy arrangement. The tremendous heat and pressure there cause the hydrogen atoms to smash into each other so hard that their electrons are knocked off. As a result, there is a sea of free-swimming electrons and free-swimming protons all mixed up.

If the temperature in the core of a star is less than about 7 million degrees, the free-swimming protons don't do much of anything except bump into each other. Their plus electrical charges act as invisible energy bumpers that prevent them from sticking together. When two protons collide they just bounce off one another, like two billiard balls. But when the temperature goes higher than 7 million degrees, the free-moving protons smash into each other hard enough so that two protons can fuse into a single lump of matter. As they do, one of the protons is changed into a *neutron*, a particle with no electric charge at all. During the change, a particle called a *positron* is created. At the same time something called a *neutrino* also is created. (Aristotle would be lost!)

Next, the positron collides with a free-swimming electron. That collision destroys both the positron and the electron and produces two gamma rays. So a tiny amount of hydrogen mass has been changed into two bursts of gamma-ray energy and one neutrino. What happens next? If you follow the next two steps in the diagram, you find that eventually the nucleus of one atom of ordinary helium (^{4}He) is built up. The energy released by all those trillions of trillions of fusion reactions going on *every* second in the Sun's core then takes about a million years to work its way up to the surface. Once in space, the energy speeds to Earth in only 8 minutes, at the speed of light.

If you got lost with all those funny-sounding particles zipping around, don't worry. The important thing to remember is that hydrogen nuclei (protons) fuse and built up helium nuclei. As they do, HUGE amounts of energy are produced. Later, the helium is used as more fuel.

So the Sun produces all its energy by burning up some of its mass as it fuses hydrogen into helium. Along the complex chain of fusion reactions energy is released. Hydrogen mass is continuously being changed into energy. And that is how virtually all stars—except for red giants and white dwarfs—produce energy and shine.

If we think about this process for a moment, we might have cause for concern. Each minute of the day and night the Sun is losing mass by converting its hydrogen fuel into energy. But as we found earlier, the Sun has enough usable mass to shine for some 10 billion years. The Sun has been shining more or less as we see it shine today for almost 5 billion years. So the Sun is halfway through its life span. It has another 5 billion years to go before burning out.

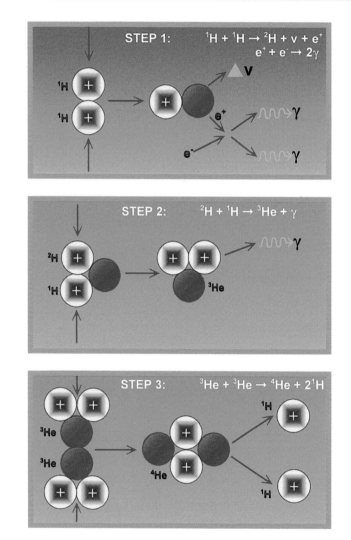

STEP 1: $^1H + {}^1H \rightarrow {}^2H + v + e^+$
$e^+ + e^- \rightarrow 2\gamma$

^1H

^1H

v

e^+

γ

e^-

γ

STEP 2: $^2H + {}^1H \rightarrow {}^3He + \gamma$

^2H

^1H

γ

^3He

STEP 3: $^3He + {}^3He \rightarrow {}^4He + 2{}^1H$

^3He

^3He

^4He

^1H

^1H

The Sun and other stars shine by smashing protons together in a fusion process that releases energy. In Step 1 two hydrogen nuclei (protons) collide and fuse. As they do, one of the protons is changed into a neutron (red ball). At the same time a neutrino (v) and positron (e^+) are created. The neutrino flies off on its own. The positron collides with an electron (e^-). Both are then changed into bursts of gamma ray energy (γ). In Step 2 the proton-neutron pair collides with a free proton. The nucleus of a light helium atom is produced along with the release of a gamma ray burst of energy. In Step 3 two light helium nuclei collide and fuse into the nucleus of an ordinary helium atom. During the fusion two protons are released and made available for more fusions.

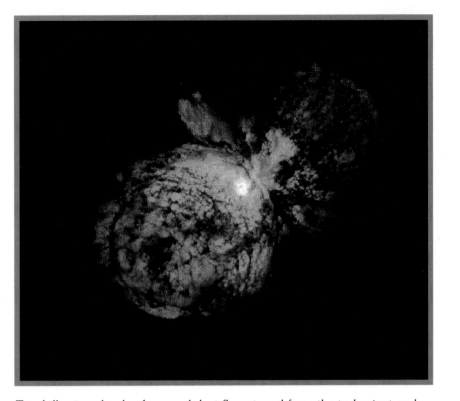

Two billowing clouds of gas and dust fly outward from the truly giant and spectacular star Eta Carinae. The clouds are puffing outward at a speed of about 1.5 million miles (2.4 million kilometers) an hour. The star itself is located between the two clouds. About 100 times more massive than the Sun, Eta Carinae gives off some five million times more energy. The outburst that produced the clouds took place 150 years ago at a distance of 8,000 light-years. Astronomers are still puzzled by Eta Carinae.

But what about other stars? Were all stars formed at the same time, and will all go out at the same time? What are those many other stars like? Most stars happen to be Sunlike stars but a bit smaller. What about all those stars out there that are not Sunlike stars? How far away are they? (Aristotle would run for cover.)

Four

How Far the Stars?

Until astronomers were able to say how far away the Sun was, they had no way of figuring out the Sun's size. It was Johannes Kepler's work in 1609 that laid the groundwork for one way of figuring out the Sun's distance. Once he worked out the exact shape of Venus's orbit around the Sun, the rest was easy. Today we can bounce radar signals off Venus to find its exact distance from us. Then by using some simple geometry, called triangulation, we get the Sun's distance. That distance turns out to be 93 million miles (150 million kilometers). We can now measure the Sun's distance very accurately by using artificial satellites put into orbit around the Sun.

Once we know the Sun's distance, it's easy to figure out its size. All we do is measure how wide the Sun appears to be at its great distance. Because we know its distance we can then calculate its actual width, or diameter. The Sun is about 865,000 miles (1,400,000 kilometers) across. That's large enough so that 110 Earths could be lined up across the Sun.

Measuring Parallax

It wasn't until 1838 that someone measured the distance to a star other than the Sun. He was the German astronomer Friedrich Bessel, and he used a method called parallax shift. It's easy to understand parallax. Hold your thumb up at arm's length and sight along it to an object way across the room, or across the street. First blink one eye and then the other. Your thumb appears to jump back and forth across the object, or across the trees out through the window. The jump is called *parallax*.

Now instead of using your thumb and trees outside your window, we'll use a nearby star (your thumb) as seen against the background of more distant stars (the trees). And instead of using the distance between your two eyes, we'll use the distance across Earth's orbit around the Sun. First we photograph the star whose distance we want to know in June. Six months later, when Earth has traveled halfway around its orbit and is on the opposite side of the Sun, we photograph the star again. When we compare the two photographs we see that the nearby star seems to have shifted its position, or jumped, against the background stars. By measuring the angle of parallax jump, we can then say how far away the nearby star is. That is just how Bessel measured the distance to a star called 61 Cygni, a star in the constellation Cygnus the Swan. Bessel's distance came very close to 61 Cygni's actual distance of 11.36 light-years. A *light-year* is the distance light travels in one year, which comes to about 6 trillion miles (10 trillion kilometers).

The nearest star to the Sun is Alpha Centauri, 4.3 light-years away. The light reaching your eyes tonight from Alpha Centauri left that star 4.3 years ago, so you are seeing that star as it was that long ago in the past. When you gaze into space you are actually

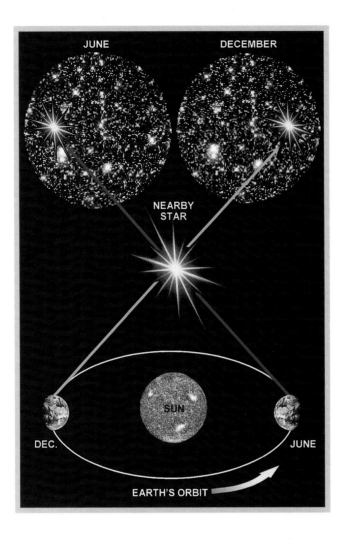

JUNE   DECEMBER

NEARBY
STAR

SUN

DEC.   JUNE

EARTH'S ORBIT

To measure a star's distance by its parallax shift: In December the astronomer photographs the star whose distance is unknown. The star appears near the right edge of the field of background stars. Six months later, in June, when Earth has traveled halfway around its orbit and is on the opposite side of the Sun, the astronomer photographs the star again. The star now appears to have shifted over to the left edge of the field of background stars. By measuring the amount of that parallax shift, the astronomer can then work out the distance to the star.

looking back in time. The Sun is 8 light-*minutes* away. The more distant stars are thousands and millions of light-years away. So when we observe the most distant stars we see the Universe as it was just after it was formed some 12 billion or so years ago!

The farther away a star is, the smaller and smaller its parallax shift becomes. Eventually the shift is too small to measure. Then how do we measure those very distant stars? Before we answer that question, it will help to know something about a special class of stars called *variable stars*. Variable stars are easy, and fun, to observe. Astronomers have known about them for many years, but we don't know as much about some of them as we would like to.

Stars That Can't Make Up Their Minds

Some of the most interesting stars to watch are the variables. We know of more than 25,000 of them. They can be thought of as stars that can't make up their minds, because they go through cycles of brightening, then dimming, and then back to bright again. The first such star studied is a giant red star named Mira, in the constellation Cetus the Whale.

Mira-type variables have periods of about 300 days. That means that it takes the star that long to complete one cycle of going from bright to dim and back to bright again. At its brightest and largest, the star is about 15 times brighter than it is when dimmest. Mira swells up to 500 times larger than the Sun. That means that 55,000 Earths could be lined up across Mira!

There are several other types of variable stars. Two types are of special interest because they have been used as yardsticks to measure the distance to galaxies beyond the Milky Way and to certain clusters of stars. One type is called RR Lyrae variables, named after the first such variable star discovered in the year 1901 in the constellation Lyra the Harp. We know of some 3,000 in the Milky Way. The stars are yellow-white giants with periods of brightness from 6 to 18 hours. A typical RR Lyrae variable may double in brightness in less than half an hour, then fade back to dim in about 4 hours.

A key thing about the RR Lyrae variables is that all of them are from 50 to 65 times brighter than the Sun. So no matter where we spot an RR Lyrae variable in the Galaxy, we know just

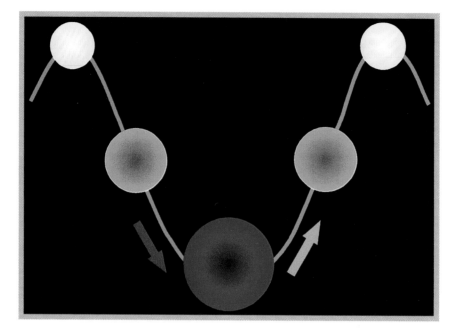

From small and bright to big and dim and back to small and bright again. That behavior marks once cycle, or period, in the life of a variable star. Some variable stars go from bright to dim and then back to bright again in only a few hours or a few weeks or more. When brightest, a typical variable may be 15 or more times brighter than when dimmest.

about how bright it is, and that is important to its usefulness as a cosmic yardstick. If we see an RR Lyrae variable that appears somewhat dim, it simply means that the star is rather far away—like dim-appearing automobile headlights shining at us from a mile or more away. If we see an RR Lyrae variable that appears bright, that must mean that the star is somewhat closer to us—like the blinding headlights of a car just about to speed past us in the opposite lane. But all the while we know that both of the stars are shining equally brightly, just as we know that the headlights of both automobiles are shining equally brightly. It is only the distance in both cases that makes the starlight and headlights *appear* brighter or dimmer. Astronomers call that effect *apparent brightness*, and it is a very important idea in astronomy. Here's why:

1) An astronomer can tell that a star is an RR Lyrae variable by measuring the star's period of brightness. 2) Knowing that the star is an RR Lyrae variable, the astronomer knows how bright the star actually is. 3) The astronomer next can use an ordinary light meter to measure the star's apparent brightness. 4) Once the apparent brightness is known, the astronomer can then use some simple arithmetic to figure out the star's distance. It is the difference between the star's actual brightness and its apparent brightness that reveals its distance.

With that method of measuring star distances, astronomy took a giant leap and the Universe grew bigger. (Aristotle would have been amazed. Ptolemy would have denied it. Galileo would have cheered!)

Before we leave the variable stars, there is one other type we should know about. They are called *Cepheid variables*, named after the first such star observed in 1784 in the northern constellation Cepheus the King. All Cepheids are hot white and yellow giants with periods that can be measured down to a fraction of a second. If only the ancients had known about these stars, and had a way to see them, they could have used the stars as fine clocks. The periods of Cepheids range from a few hours up to about 50 days, but most are from 5 to 8 days.

In 1912 the American astronomer Henrietta Leavitt, of Harvard University, did pioneering work that led other astronomers to use Cepheid variables to measure star distances even greater than distances measured by RR Lyrae variables. She found that a Cepheid's true brightness was directly related to its period of brightening and dimming. The reason the Cepheids turned out to be a better yardstick was that they are giant and very bright stars that can be seen at far greater distances than the RR Lyrae variables. While the RR Lyrae variables were just fine for measuring distances within the Milky Way galaxy, their fires were too feeble to be seen in faraway galaxies beyond our own. The Cepheids, however, burned brightly from those greater distances.

The Andromeda Galaxy (M31) was once thought to be a mere fizzy patch of dust within our home galaxy. But astronomer Edwin Hubble showed it to be a galaxy far beyond our own. Variable stars called Cepheid variables provided the cosmic yardstick Hubble needed to estimate the galaxy's distance of some 2.3 million light-years. From a distance, our home galaxy, the Milky Way, would look something like Andromeda.

So it was to the supergiant Cepheid variables that the American astronomer Edwin Hubble turned to measure the distance to the first star system beyond our own—the Andromeda Galaxy, also known as M31. Hubble's discovery actually settled a debate about the nature of distant fuzzy patches of light that no one could be sure about. Grounds for the debate were set way back in 1750 when the English astronomer Thomas Wright said that our home galaxy of stars, the Milky Way, did not stretch away forever but had boundaries. He also felt that those fuzzy patches were collections of stars far beyond the Milky Way. He was not alone in that belief. The German philosopher Immanuel Kant agreed with Wright, calling those nebulae "island universes." But around 1920 the American astronomer Harlow Shapley disagreed. He believed that the fuzzy patches were nearby and part of the Milky Way. And here is where the Cepheid variables came in.

The year was 1924 and the place was California's 100-inch (2.5-meter) telescope on Mount Wilson. Hubble was able to identify Cepheid variable stars in M31. Then by comparing their apparent brightness with their known actual brightness, he was able to say that the Andromeda Galaxy lay at the great distance of nearly a million light-years. But Hubble's distance was way off. The reason was that there were actually two different types of Cepheid variable stars, one brighter than the other. Hubble was not aware of that important difference. When a correction was made, the Andromeda Galaxy turned out to be 2.3 million light-years away. So Wright and Kant had been correct. M31 and those other nebulae with spiral shapes were cities of stars like our own galaxy but vastly distant. And it was the light from one class of variable stars that provided the key to the puzzle.

But other than variable stars, what other kinds of stars wink down on us from the eternity of the heavens?

Five

Stars Galore

Stars That Go Boom! in the Night

When he was observing the sky on the night of November 11, 1572, the Danish astronomer Tycho Brahe couldn't believe his eyes when he saw a sign that all was not right in the heavens, at least not right according to Aristotle.

Aristotle had said that beyond the distance of the Moon the heavens do not change. Tycho saw that they did, and that bothered him. A star in the constellation Cassiopeia the Queen suddenly flared up as what today we call a *supernova*. A supernova

Tycho Brahe was a keen observer and master builder of instruments for measuring the positions of the stars and planets. Here he is shown in his observatory with his numerous assistants. King Frederick of Denmark built the elaborate observatory for him on the island of Hven. Tycho lived from 1546 to 1601. Had he lived another nine years, he would have been astounded by what Galileo saw when he became the first person to study the heavens with a telescope.

is a star that explodes itself to bits and so appears very bright to an Earth observer. In only a few hours the star may swell up in brightness until it becomes one of the brightest stars in the entire sky. "Tycho's Star," as it came to be called, was nearly as bright as Venus. Then over a period of two years it gradually faded from view. (Aristotle would not have liked supernovas.)

On July 4 in the year 1054 American Indians and Chinese astronomers made records of a rare event they saw in the sky. A star exploded itself to bits. The Chinese called it a "guest star," and said it "was visible in the day like Venus, with pointed rays in all four directions. The color was reddish white . . . It was seen altogether for twenty-three days." Drawings of what seem to have been that rare event have been found in northern Arizona on a wall of Navajo Canyon and in a cave at White Mesa.

The Chinese records give us a pretty good idea of what people of the time must have seen. When the star exploded, it shone with the light of 400 million Suns for a few weeks. At a distance of 6,300 light-years, the explosion sent a cloud of gases speeding outward at about 1,000 miles (1,600 kilometers) a second. We see the cloud today as the Crab Nebula in the constellation Taurus the Bull. It is still spreading outward and is now nearly 10 light-years across. Supernova star explosions such as this are among the most violent events we can observe in the Universe. Tycho's Star seems to have been a supernova. To date, only

The Crab Nebula is the gaseous remains of the "guest star" of the year 1054. It was a supernova that continues to this day to spread its exploded gases across space at a speed of 1,000 miles (1,600 kilometers) a second. The cloud is about 10 light-years across. The nebula is located in the constellation Taurus the Bull.

On July 4 in the year 1054 Chinese astronomers described the appearance of a supernova explosion, which they named the "guest star" and reported to be visible in the day like Venus. What appear to be drawings of the event also seem to have been made by American Indians of Arizona. The exploding star is shown at lower left with a crescent moon to the right.

about 6 of these catastrophic explosions have been observed. Not one has been seen in our home galaxy since Galileo first pointed his telescope skyward.

The type of supernova star we have been talking about so far are called Type II supernovas. They occur only in those Class O and Class B huge, bright, and extremely hot stars known as blue *supergiants*. They are the most massive stars known, with masses 6 or more times that of the Sun. When a Type II supernova exhausts its fuel supply in its core, the core cools down, the pressure drops, and the outer gas layers come crashing down into the core with terrible force. In less than a second, the star explodes and splashes untold tons of matter off into space. The matter then floats among the stars for billions of years as colorful clouds. Some of it is swept up and becomes part of other stars; some of it goes into new star formation and so is eternally recycled. All that remains of a star that has been blown apart as a supernova is the intensely hot and exposed core. Over billions of years the core cools, dims, and eventually becomes invisible. To this day we can still see the core remains of the Guest Star of 1054.

Less explosive stars called *nova*, or "new," stars also flash into prominence every now and then and so would upset Aristotle by causing the heavens to change. A nova is not a new star at all but one that, for some reason we don't fully understand, blows off its outer shells of gases. Novas seem to get their explosive energy as their gravity pulls away hydrogen surface gases from a nearby companion star. As the newly captured hydrogen gas builds up on the nova star's surface, it keeps getting hotter and denser until it blows up like a colossal hydrogen bomb. The explosion may make the nova increase in brightness by a million times in only a day. The nova may then fade away in only several days or weeks.

Astronomers used to think that a nova star cast off a smooth shell of gases, but in 1997 the Hubble Space Telescope produced a surprise. Nova T Pxyidis exploded off thousands of gaseous blobs, each the size of the Solar System. T Pxyidis is now thought to recharge and repeat its nova cycle once every 20 years or so.

Oddball Stars Unlimited

There are other kinds of oddball stars that fascinate and baffle astronomers. Among them are the *eruptive variables*, such as the one in the constellation Cygnus the Swan and named SS Cygni.

Most of these stars belong to *double-star* systems—two stars relatively close and that revolve around one another. Most nova stars also seem to be members of double-star systems. But the eruptive variables are tame compared to nova stars. They are bluish dwarf stars that swell up to a hundred or so times brighter than usual. Then after a few days they slowly fade back to their usual brightness.

SS Cygni stars, as this group of eruptive variables are called, may fuel themselves much the same way nova stars do. They pull off the outer layers of hydrogen from their companion stars. When an eruptive variable has collected enough extra surface hydrogen, the gas erupts and puffs up the star. Although these stars are less violent than nova stars, some astronomers think they may one day evolve into full-scale novas.

Feathery wisps of gases float through space as cast-off remains of a supernova explosion. These gases contain heavy elements, including iron and uranium, for example. The red gases are charged atoms of sulfur, the blue of oxygen, and the green of hydrogen. The cloud was blasted away by a supernova that exploded 15,000 years ago and is some 2,500 light-years away. It is part of a nebula named the Cygnus Loop seen in the constellation Cygnus the Swan.

Another group of oddball stars are called *flash stars*. The Pleiades cluster in the constellation Taurus the Bull has several. After flashing into brightness, the star may remain bright from a few minutes up to 3 hours. It's possible that these stars are young stars whose core nuclear fires have not yet begun to burn steadily. Instead, nuclear fusions sputter on and off like a stalled car being encouraged to start up again.

Binoculars or a small telescope will bring into view a fascinating star with the Arabic name Al Shilyak, meaning "the tortoise." Its formal name is Beta Lyrae, and it is the second brightest star in the constellation Lyra the Harp. Discovered in 1784, it is another double-star system in which one star eclipses the other every 13 days as the two revolve about each other. The major

The Pleiades open cluster of stars is in the winter sky constellation Taurus the Bull. The stars, also called the Seven Sisters, are young giant stars shining with a bluish-white light. The pale blue glow is caused by gas and dust left over from the original gas cloud that gave birth to the stars. A number of variable stars called flash stars are among the Pleiades group. They flare up and dim over periods of a few minutes to a few hours, as if sputtering in an attempt to get their nuclear furnaces going steadily.

star in the pair is a giant almost 20 times larger and 3,000 times brighter than the Sun. The smaller companion is about 15 times the Sun's size. Beta Lyrae is one of the largest known double-star systems. But that is not what makes the pair so interesting.

It is their closeness—only 22 million miles (35 million kilometers)—that makes them so fascinating. Only about Earth's distance from Venus, the two-star strong gravitational tug on one another stretches each star out into an egg shape. Gravity also causes the more massive star of the pair to pull surface gases away from the other star. Adding to this gravitational display, gases are cast off by both stars in the form of a huge stream of matter that spirals outward around the two stars. Since both stars keep losing matter to space, it's hard to know what the future will hold for them.

Perhaps the most interesting thing about all of these oddball stars is that we know so little about what makes them work. Astronomers will continue to study them for many years to come.

The Three Major Star Types

Ever since astronomers have been observing the behavior of stars and photographing their spectral-line fingerprints, they have learned that, like people, every star in the heavens is unique. Even so, we can group most stars into three major types—very massive giant stars, medium-mass stars like the Sun, and low-mass dwarf stars.

The stars form out of those unimaginably large clouds of gas and dust called *nebulae*. We can observe many nebulae in our home galaxy and in galaxies far beyond. A star begins as a collection of gas and dust in the form of a globule. A globule grows larger and more massive as it keeps sweeping up more gas and dust. Gravity of the globule packs this dark and cold star matter ever tighter into a sphere-shaped object. The collecting and packing goes on for millions of years, heating the accumulating gas and dust as the globule grows. Eventually it heats up enough to glow a dull red. The globule has become a *protostar*, meaning "early star." Eventually, the temperature and pressure of hydrogen packed into the core region reaches 7 million degrees.

When it does, the protostar flashes into brilliance as a new, fully-formed star. Its core has become a nuclear furnace fusing hydrogen into helium.

The kind of star a new star becomes depends entirely on how much gas and dust a protostar manages to pack into itself as it forms. "Mass" is the key word here. Lots of mass builds those

enormous and extremely hot giant and supergiant stars like Rigel and Sirius. They burn with a hot bluish-white light. A medium amount of mass leads to "average" Sunlike stars that burn with a cooler yellowish light. And a low mass protostar evolves into those still cooler and smaller stars called red dwarfs.

The more mass a star collects, the hotter it becomes and the brighter it shines. But the brighter it shines, the faster it uses up its hydrogen fuel supply in its core region. The blue supergiants have the shortest life spans of all stars. They last only a few hundreds of millions of years. Because Sunlike stars shine with a cooler

Star birth galore goes on in the galaxy NGC 4214, some 13 million light-years away. The youngest stars are still encased in their cocoons of hot gases, seen as five large white patches at lower right. At center is a cluster of hundreds of massive blue stars 10,000 times brighter than the Sun. The faint red dots are ancient stars that show that star formation has been going on in the galaxy for billions of years.

light, they use their fuel more slowly and so have longer life spans of about 10 billion years. The still cooler red dwarf stars have the lowest core temperatures, barely high enough to keep their nuclear furnaces burning. As a result, they have the longest life spans, measured perhaps in a trillion or more years. The nearby red-dwarf star Beta Centauri has only about one-tenth the mass of the Sun and is only half the size of Jupiter! It would take 13,000 stars like it to equal the energy output of the Sun.

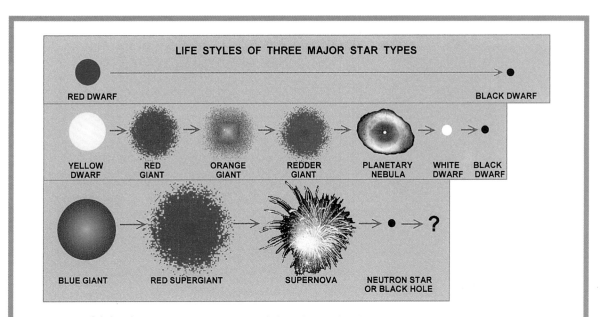

LIFE STYLES OF THREE MAJOR STAR TYPES

RED DWARF   BLACK DWARF

YELLOW DWARF → RED GIANT → ORANGE GIANT → REDDER GIANT → PLANETARY NEBULA → WHITE DWARF → BLACK DWARF

BLUE GIANT → RED SUPERGIANT → SUPERNOVA → NEUTRON STAR OR BLACK HOLE → ?

Of the three major star types, red dwarfs are the dimmest and have the longest life spans of about a trillion years. As they exhaust their hydrogen fuel they simply become dimmer until they "go out" and remain as black dwarfs. Sunlike stars have shorter life spans because they burn their hydrogen fuel faster. As their fuel runs low, they first swell up as a red giant, blow off some of their gases as a planetary nebula, then shrink as a white dwarf, and finally cool as black dwarfs. The massive blue giants have the shortest life spans. They eventually swell up as supergiants, explode as supernovas, and then end up either as neutron stars or black holes.

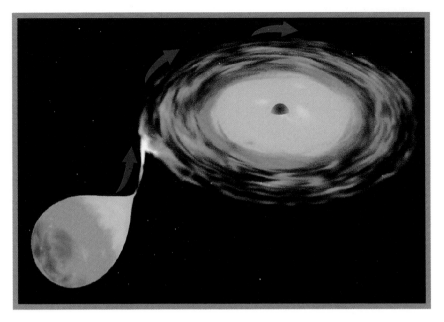

How do we know if a black hole is there if we can't see one? A black hole's super strong gravity pulls gases off a nearby star with such great force that the gases give off Xrays as they form an accretion disk of matter that spirals into the black hole.

Earlier you found that it is only the blue supergiant stars that go on rampages as a supernova. Bigger is not always better, at least if you want to live long. Depending on just how much mass a blue supergiant star has, it may end its life as those exotic objects called *black holes*. A black-hole star may be 10 times more massive than the Sun. The matter left in the core of an exploded supernova may be crunched together so forcefully that it becomes super dense—so dense that the object's gravity prevents any energy from escaping, including light. One astronomer has described a black hole as an object that dug a hole, jumped in, and then pulled the hole in after itself!

Sunlike stars end their lives much less dramatically, at least in the end. When they run out of nuclear fuel they collapse in on themselves. The massive infall of matter sends the core temperature soaring, and the star puffs up as a red giant star. That's the beginning of the end. Most of the puffed-out gases then tumble back into the star and may repeat the drama a second time. Much of the bloated outer gas layers escape as a gas cloud called a *planetary nebula*. But eventually most of the gases just settle back into the core region. With the nuclear furnace shut down for good, the gases get packed so tightly in the core that the star becomes a *white dwarf*, a tiny star that shines with an intense white light.

Interestingly, a tiny white dwarf is capable of swelling up and becoming a Type I supernova. A white dwarf that has a nearby companion may gravitationally keep pulling away some of the companion stars' outer gases and so grow more massive. When its mass becomes as much as 1.44 times more massive than the Sun, the star erupts violently as a Type I supernova.

Over billions of years after entering white dwarf land, a star simply cools down until it fades from view as a *black dwarf*. The Universe probably is too young for there to be black dwarfs out there yet since their cooling time is so very, very long. Red dwarf stars do not go through the red giant stage. When their fuel is used up they just collapse into a black dwarf.

In the 1990s astronomers using the Hubble Space Telescope discovered a kind of star never before seen. Called *brown dwarfs*, they existed only in theory. They have so little mass that their cores are unable to heat up enough to start fusing hydrogen into helium. As a result, they shine only by the crushing force of their gases by gravity, and they glow with a cool, brownish-red light even cooler than the red dwarfs. There must be millions of these near-miss stars out there shining so faintly that we will spot one only rarely, at least until some new method of observing them comes along. One such brown dwarf, named 8B, was spotted by Hubble only because it's so close, only 20 light-years away.

The End and New Beginnings

So in the great cosmic scheme of things, stars are born, shine for a cosmic while, and then go out, some with a whimper, others with a bang. The matter cast off by exploding and dying stars is not lost. It does not just go away. It sweeps across interstellar space and mixes with a nebula here, another nebula there. So cast-off stellar matter is caught up in the cosmic recycling machine and eventually becomes the gas and dust that go into new star and planet formation. The iron, carbon, and other heavy elements in your body were once part of a massive blue supergiant star and helped that star shine for a while. When it exploded and cast off those heavy elements as a cloud, the cloud eventually mixed with the nebula out of which the Sun and its nine planets were formed.

So you, I, and all living matter on Earth contain pieces of a star that long ago helped light up the Universe.

The story of astronomy tells us that there is always something new beyond the Sun. Each month, each week, astronomers speak of their amazement and awe when their powerful telescopes and satellite observatories reveal new wonders among the stars. How interesting it would be if we could turn the clock ahead 500 years to find out what astronomers of that future time had learned and what they think of our ideas about the stars today. Aristotle, Ptolemy, Tycho, and Galileo would all agree.

Glossary

Apparent brightness—the observed brightness of a light source, depending on its distance from the observer. The more distant a light source, the dimmer it appears.

Astrology—the superstitious belief that the stars and planets affect future courses of a person's life based on their positions in the sky at the time of the person's birth.

Black dwarf—a burned-out star that has passed through the white dwarf stage and has dimmed to the extent that it is no longer observable.

Black hole—an incredibly dense and massive star that has burned itself out. Black holes are thought to be so dense that radiation is unable to escape from them; hence, we cannot observe them directly.

Brown dwarf—a very low-mass star that is not hot enough to start up fusions of hydrogen into helium and, as a result, is a very dim radiator of energy.

Cepheid variable—a hot white and yellow giant variable star with a period ranging from a few hours to about 50 days. These stars are used as cosmic yardsticks to measure distances to galaxies beyond the Milky Way.

Continuous spectrum—a spectrum smoothly graded from red to violet with no breaks or gaps or lines.

Dark-adapted—when one's eyes have become used to the dark and are able to function at their maximum, usually over a period of about 10 minutes.

Double-star system—two stars relatively close to one another and that revolve about each other.

Electromagnetic spectrum—the entire range of radiation from gamma rays at one extreme to radio waves at the other extreme.

Electron—a negative unit of electricity occurring at various energy levels of all atoms.

Eruptive variable—a bluish dwarf star that gravitationally draws off hydrogen from the surface gases of a companion star. Eventually the accumulation of hydrogen flares up, temporarily increasing the eruptive variable's brightness. These stars may eventually evolve into full-fledged nova stars.

Flash star—a star that gravitationally draws off hydrogen from a companion star and then burns the accumulated hydrogen at its surface in outbursts that temporarily increase the brightness of the star.

Light-year—the distance that light travels in one year at the rate of 186,282 miles (299,792 kilometers) per second. That distance amounts to about 6 trillion miles (10 trillion kilometers).

Nebula—a vast cloud of gas and dust, many of which are seen in our home galaxy and in galaxies far beyond our own.

Neutrino—a subatomic particle with little or no mass, capable of passing through the entire planet as if it weren't even there. Neutrinos continue to be regarded as mystery particles.

Neutron—particles lacking an electric charge and present in all atoms except hydrogen. Neutrons have approximately the same mass as protons.

Nova—a so-called "new" star. Nova stars flare up in brightness explosively as a result of drawing off hydrogen gas from a nearby companion star. When enough excess hydrogen accumulates on a nova, it erupts violently.

Nucleus—the central cluster of protons and neutrons in atoms.

Parallax—the apparent shift in position of a nearby star among the more distant background stars when the nearby star is observed from two different positions.

Planetary nebula—a nebula such as the Ring Nebula in Lyra, once mistakenly thought to be a planet. The faint greenish color of planetary nebulae gives them the appearance of the planet Uranus.

Positron—a positively-charged electron.

Proton—one or more positively-charged particles found in the nucleus of all atoms.

Proton-proton chain—a complex series of nuclear reactions occurring in the cores of stars during which hydrogen nuclei (protons) fuse and build up the nuclei of helium atoms.

Protostar—a star in its early stages of formation before it has reached high enough temperatures to set off nuclear reactions in its core.

Rapid oxidation—the rapid combining of oxygen with another substance; burning and spontaneous combustion are caused by rapid oxidation.

Red dwarf—a low-mass star whose core temperature barely reaches a temperature high enough to initiate the nuclear fusion of hydrogen into helium.

Spectroscope—an instrument that can be attached to a telescope and used to analyze the light of a star in order to determine the chemical makeup of the star; also to determine the motion of the star toward or away from the observer.

Stars—enormous globes of hot gases that generate energy and so shine by emitting that energy over very long periods of time. The Sun is a star.

Supergiant—an extremely large star, such as the blue supergiant star Rigel, that is extremely hot and has a relatively short life span; also an extremely large star, such as the red supergiant Betelgeuse, which has reached the end of its life span.

Supernova—a very massive giant star that is no longer able to fuse atomic nuclei and, as a result, collapses and explodes violently.

Variable star—any star that exhibits cycles of brightening and dimming.

Visible spectrum—that part of the electromagnetic spectrum ranging from violet to red and whose radiation is visible to the eye.

Wavelength—the distance between two successive crests of an ocean wave, or a wave of radiation along the electromagnetic spectrum.

White dwarf—a small star representing the last stage in the life span of Sunlike and red dwarf stars.

Ziggurat—observation platforms built by Babylonian astronomer-priests several thousand years ago.

Benestad, Rasmus E. "Solar Activity and Global Sea-surface Temperatures." *Astronomy & Geophysics*, pp. 14–17, June 1999.

Bernstein, Max P., Scott A. Sandford, and Louis J. Allamandola. "Life's Far-Flung Raw Materials." *Scientific American*, pp. 42–49, July 1999.

Krupp, E.C. *Skywatchers, Shamans & Kings*. New York: John Wiley & Sons, Inc., 1997.

Needham, Joseph (abridged by Colin A. Ronan). *Science & Civilisation in China: 1*. Cambridge: Cambridge University Press, 1978.

Gallant, Roy A. *Astrology: Sense or Nonsense*. Garden City, NY: Doubleday & Company, Inc., 1974.

———. *Earth's Place in Space*. Tarrytown, NY: Benchmark Books, Marshall Cavendish, 2000.

———. *When the Sun Dies*. Tarrytown, NY: Marshall Cavendish, 1998.

———. *How Life Began: Creation Versus Evolution*. New York: Four Winds Press, 1975.

———. *Beyond Earth: The Search for Extraterrestrial Life*. New York: Four Winds Press, 1977.

———. *Private Lives of the Stars*. New York: Macmillan Publishing Company, 1986.

Kaler, James B. *Astronomy!* New York: Addison Wesley, 1997.

Reed, George. *Eyes on the Universe*. Tarrytown, NY: Benchmark Books, Marshall Cavendish, 2001.

Ronan, Colin A. *The Astronomers*. London: Evans Brothers Limited, 1964.

Index

Page numbers for illustrations are in **boldface**.